CHEMISTRY QUICKIES

Vivian W. Owens

ESCHAR PUBLICATIONS
Waynesboro, Virginia

CHEMISTRY QUICKIES

By Vivian W. Owens

Published by:
ESCHAR PUBLICATIONS
P.O. Box 1196
Waynesboro, VA 22980

First Printing 1998
Printed in the United States of America

Library of Congress Catalog Card Number:
97-77828
ISBN: 0-9623839-7-x

You may order single copies prepaid direct from the publisher for $13.95 plus $2 for postage and handling. (Virginia residents add 4.5% for state sales tax). For terms on volume quantities, please contact the publisher.

DEDICATION

To my Chemistry students, for showing me how much fun quick learning can offer to anyone.

ACKNOWLEDGEMENTS

I wish to thank the students at
Wilson Memorial High School
and at
Stuarts Draft High School
who tested the contents of this book.
I am also grateful to
DuPont chemist, Portia Bass,
and to Ken Patterson,
fellow chemistry teacher,
for their valuable time and expertise
in reading the manuscript for this book
and offering criticisms and suggestions.

By Vivian W. Owens

Juvenile Fiction

I MET A GREAT LADY
I MET A GREAT MAN
THE ROSEBUSH WITCH
NADANDA, THE WORDMAKER

Parent Helpers

CREATE A MATH ENVIRONMENT
PARENTING FOR EDUCATION

General Audience

CHEMISTRY QUICKIES

CHEMISTRY QUICKIES

CONTENTS

SCIENTIST ILLUSTRATIONS

INTRODUCTION

Everybody likes a little fun. People play games almost anywhere, because there's something in the human spirit that enjoys the hint of challenge. Throughout my years of teaching Chemistry, I've found that the students in my classes are just like everybody else, anywhere else. They like fun--particularly, fun with chemistry. Realizing this, I sought ways to liven their interest and to motivate them by creating learning mediums using simple fare of mind and reading. In this way, CHEMISTRY QUICKIES was born.

Rising from my desire to jump-start the day with something fun and tantalizing which could engage all students simultaneously, CHEMISTRY QUICKIES took the form of questions requiring less than five minutes of textbook research. Appropriately used, they can increase a student's critical reading skills, while also increasing critical thinking skills. If a teacher needs to review test concepts in a light manner or broaden development of chemical principles in a non-hammering method, this allows you. Problem solving gains a friendly presence in an atmosphere deliberately rich in eye-blink fun.

Possibilities for QUICKIES are exciting. They permit students to stretch academic muscles before beginning the required duties of the day, thus shaping a good frame of mind. Quick recall may be evaluated or notes may be consulted.

CHEMISTRY QUICKIES

INTRODUCTION

Some would say many of these questions touch the familiar in us and around us, inviting us to realize how Chemistry is common ground for living matters which concern us all. Whether our attention is focused on food, medicine, hobby, or the environment, we peer curiously into Chemical literature, searching for answers. Some answers we know already due to prior exposure.

On the other hand, if you are reading CHEMISTRY QUICKIES in a non-academic environment, you have no need to consult a textbook, since the answers are given to you in the reference section. In all respects, the intent of this book is quick, gamelike response, and secondarily, it is a resource book. From this angle, you may use it alone or with a partner or team.

Obviously, teachers, students, home schooling parents, and tutors will enjoy CHEMISTRY QUICKIES, but something is here for anyone from any walk of life. Something in each of us moves toward a challenge, expecially when that challenge is fun and quick.

Readers, I sincerely want you to have as much pleasure reading this book as I had writing it.

Vivian Owens

HOW TO USE CHEMISTRY QUICKIES

CHEMISTRY QUICKIES is not a textbook, but it can be used as a supplement with chemistry textbooks, because it presents a questioning approach to critical reading and critical thinking. Since it was arranged for daily sequence without dependence on topic grouping,students acquaint themselves with their textbooks through deliberate exploration that is much faster than a normal course development schedule. In this friendly pursuit of information, students peek at concepts sometimes not covered, and they also ponder over ideas which may prompt them to broaden their interest in Chemistry.

Listed below are some suggestions for using **CHEMISTRY QUICKIES** for instructional purposes.

- Study the "Reference Guide To Topic Items" when you want to preview specific concept items.
- Introduce new concepts.
- Use CHEMISTRY QUICKIES daily to jump-start a class period.
- If you are a classroom teacher, consider awarding points to students who correctly answer questions. These points may add up to a replacement grade for a quiz or test by the end of a marking period.
- In a game format, challenge students through fun, competitive matches.
- Treat CHEMISTRY QUICKIES as one method of quick learning when there is not time to thoroughly devour a subject.

HOW TO USE CHEMISTRY QUICKIES

- For a Block Schedule of one semester duration, consider CHEMISTRY QUICKIES as one approach to exposing students to a broad range of concepts when knowledge depth is an unreasonable goal.
- Review for tests or quizzes.
- If you want to provide a short lesson of singular, focused interest, do this with a CHEMISTRY QUICKIE. For example, you may ask a "Quickie" about the green coating (patina) on the Statue of Liberty, to be followed by a twenty minute discussion on oxidation-reduction or a discussion on corrosion of metals. A short lab or short activity may accompany this focus.
- If you are a student, refresh yourself by browsing through this book. Attempt to recall ideas and link together concepts in a meaningful way.

For Non Instructional Purposes

- Choose QUICKIES to springboard your conversations.
- Play games with family or friends.
- Reacquaint yourself with Chemistry in this nonthreatening book environment.
- Read a Quickie per day just to give you food for thought.

Finally...

Use this book for fun!
HAVE FUN! FUN! FUN!

CHEMISTRY QUICKIES

CHEMISTRY QUICKIES

#1

A chemical change involves a change in the composition of a substance. True or false?

#2

Elements cannot be chemically decomposed into simpler new substances. True or false?

#3

Have you ever looked closely at the cap of the Washington monument? Of what metal is the cap made?

CHEMISTRY QUICKIES

#4

Hiccups are a mechanism used by the body to get rid of a certain gas, when the gas is in excess. Name that gas.

#5

What pH do humans need for their blood?

#6

How does the blood maintain its necessary pH?

CHEMISTRY QUICKIES

#7

What is the formula for calcium chloride?

#8

Name the common household liquid, found in many kitchen cabinets, that is 5% aqueous solution of acetic acid, made by fermenting fruit or grain.

#9

What is the common name for calcium oxide?

CHEMISTRY QUICKIES

#10

What property of water causes sidewalks to crack and frozen water pipes to break?

#11

What is the formula mass in grams for sodium hydroxide?

#12

What is the symbol for the element, bromine?

CHEMISTRY QUICKIES

#13

What is the symbol for the element, americium?

#14

If you were hot-air ballooning, how would you control the height of the balloon over the ground?

#15

Myricyl alcohol, $C_{30}H_{61}OH$, reacts with palmitic acid, $CH_3(CH_2)_{14}COOH$, to form the ester, Beeswax. What is the formula for beeswax?

CHEMISTRY QUICKIES

#16

What single term expresses the quantity of moles per volume?

#17

Explain transmutation.

#18

Dry ice is a very useful refrigerant. Is dry ice a form of (a) hydrogen (b) carbon dioxide or (c) hydrogen sulfide?

CHEMISTRY QUICKIES

#19

Does the Periodic Table contain more nonmetals or metals?

#20

By how many degrees does the temperature change as ice melts?

#21

When water is added to calcium oxide, calcium hydroxide is formed. What is the common name for calcium hydroxide?

CHEMISTRY QUICKIES

#22

What does HBO have to do with chemistry or oxygen?

#23

When water is softened, a resin is used. Which cations in hard water are attracted to the resin?

#24

What is "frothy" water?

CHEMISTRY QUICKIES

#25

George Washington Carver was a great American scientist who created over 350 products from the peanut. He taught southern farmers how to save their soil. Name the 3 universities with which Dr. Carver was associated.

#26

What is meant by "complete protein"?

#27

What are elastomers?

#28

Can two isotopes of the same element have different atomic numbers?

CHEMISTRY QUICKIES

#29

Give an ordinary, every day example of an elastomer.

#30

Where are the two U.S. vitrification plants located?

#31

Why are materials such as boron and cadmium placed in control rods of nuclear reactors?

CHEMISTRY QUICKIES

#32

Write the formula for ammonium chloride.

#33

You run an experiment to test water. Iron (III) Thiocyanate is produced. How will you know it?

#34

From what is paper made?

CHEMISTRY QUICKIES

#35

Give a name to a molecule composed of only two atoms.

#36

How many different elements make up a binary compound?

#37

CFC is a major contributor to the depletion of the ozone layer. What are CFC's?

#38

Rhodopsin is one of the molecules responsible for vision. It contains two parts, retinal and opsin. Retinal occurs in several isometric forms, and in the 11-cis form it is bound to opsin but separates from it when rhodopsin absorbs light. What isometric form does retinal take upon this separation?

#39

What do yeast cells contain that causes them to act as catalysts?

#40

Sodium and potassium are essential to the proper functioning of which body system?

CHEMISTRY QUICKIES

#41

Name this chemical: Historically, it was widely traded and heavily taxed. In the Mediterranean and Middle Eastern countries, it was highly valued. Seas and lakes taste a certain way because of it. Brine solutions contain it.

#42

How do the large bubbles in bubble gum form? This does not occur in regular gum.

#43

What gives sourdough bread its unique taste?

CHEMISTRY QUICKIES

#44

Since unsaturated fats are better for your health than saturated fats, why do food manufacturers hydrogenate oils in order to reduce unsaturation?

#45

A violet colored mineral, often used in jewelry, is formed when silicon dioxide, SiO_2, receives a trace of iron.

What is this mineral?

#46

What is the chemical name for China clay or kaolin?

CHEMISTRY QUICKIES

#47

Clay is composed of what elements?

#48

What is another name for a basic solution?

#49

Vitamin C is a water soluble vitamin. What is another name for Vitamin C?

CHEMISTRY QUICKIES

#50

What is the atomic mass for cesium?

#51

The horizontal row on the Periodic Table is called ____________________? The vertical column is called _____________________?

#52

Give another name for the helium-family elements.

CHEMISTRY QUICKIES

#53

What is the name for biological catalysts?

#54

A kind of solution contains a higher concentration of solute than a saturated solution at the given temperature. Name the kind of solution.

#55

When is water considered softened?

CHEMISTRY QUICKIES

#56

MARIE CURIE
1867-1934

Marie, along with her husband, Pierre, discovered the radioactive elements, radium and polonium. They received the Nobel Prize in Physics with Henri Becquerel in 1903. Additionally, Marie received another Nobel Prize. What year was this?

CHEMISTRY QUICKIES

#57

Clover and alfalfa are examples of legumes. What element is involved in a fixation process that allows legumes to fix bacteria in their roots?

#58

What is the major protein in egg whites?

#59

The protein in egg whites is digested in the small intestine by pepsin. Is pepsin an acid or an enzyme?

CHEMISTRY QUICKIES

#60

Name the world's largest vitrification facility.

#61

Name this compound:

```
   H H H H H H
   | | | | | |
 H-C-C-C-C-C-C-H
   | | | | | |
   H H H H H H
```

#62

Is frozen smoke chemically inert?

CHEMISTRY QUICKIES

#63

Amino acids are small molecules that build polymers. What polymers do they build?

#64

A deficiency of this mineral would result in poor memory, thirst, headache, appetite loss, and weakness. What foods would provide this mineral?

#65

What gas do humans give off as a waste product but plants need it for life?

CHEMISTRY QUICKIES

#66

By alpha emission, Rn-222 decays into which element (isotope)?

#67

At what blood alcohol level does the probability of causing a traffic crash double?

#68

The Haber Process is a less expensive process for manufacturing what compound?

CHEMISTRY QUICKIES

#69

In what type of food are partially hydrogenated molecules found?

#70

Uranium metal is shipped in quantities of 167 kilograms. This quantity is given a special name in the nuclear industry. What is that special name?

#71

What is the normal boiling point of water?

CHEMISTRY QUICKIES

#72

In an automobile battery, the lead sulfate formed can cause problems, if the battery is used too long without recharging. How does the $PbSO_4$ cause problems?

#73

What element gives colors to emeralds and rubies?

#74

The heat of vaporization is the energy needed to vaporize 1 gram of liquid to gas. What is the heat of vaporization for water?

CHEMISTRY QUICKIES

#75

What is the difference between jams and jellies?

#76

Why is the density of ice less than water?

#77

Name the acid some ants inject into their victims?

CHEMISTRY QUICKIES

#78

Most of the refined starch in the United States is produced from what substance?

#79

Convert 826 picoseconds into seconds.

#80

During a vigorous exercise, your body may accumulate so much acid that muscle soreness results. Which acid is that?

#81

Balance this equation:
$Pb(CH_3COO)_2(aq)+K_2CrO_4 \rightarrow PbCrO_4(Cr)+$
KCH_3COO (aq)

#82

Write the structural formula for this compound, if the molecular formula is C_2H_2.

#83

Your stomach contains gastric juice, and this gastric juice contains a strong acid. Name the strong acid found in gastric juices.

CHEMISTRY QUICKIES

#84

FRIEDRICH WOHLER
1800-1882

Wohler encouraged other chemists to synthesize organic compounds, after he synthesized urea. From what compound did he synthesize urea?

CHEMISTRY QUICKIES

#85

Buffer solutions control pH. How?

#86

Name the scientist who devised a method for reducing aluminum using electricity.

#87

What percent of proteins can the body store?

CHEMISTRY QUICKIES

#88

Is it correct to add water to a large amount of acid?

#89

A cup of tea may be too hot to drink. If the top surface is removed, it will likely then be drinkable. Why?

#90

Will a can of beverage cool just as fast in the regular part of the refrigerator as it will in the freezer compartment?

CHEMISTRY QUICKIES

#91

Why do pigs wallow in the mud to cool themselves?

#92

Name the product formed when an alcohol reacts with an acid.

CHEMISTRY QUICKIES

#93

Name the ester responsible for the flavor of oranges.

#94

Under what circumstance will oxygen carry an oxidation number of -1? Usually it exists with an oxidation number of -2.

#95

Why is it that white phosphorus must be kept and handled under water?

CHEMISTRY QUICKIES

#96

Name the ester responsible for the flavor of bananas.

#97

Name the ester responsible for the flavor of apples.

#98

Chlorine exists in two forms. One has 17 protons. One has 18 neutrons. What is different for these two forms?

CHEMISTRY QUICKIES

#99

For what does STP stand?

#100

If benzene is ingested into the body, it undergoes a chemical reaction in the liver that makes it less toxic and it will also be soluble in water. To what is benzene changed?

#101

Give the name for the gases that show very little or no deviation from the gas laws.

CHEMISTRY QUICKIES

#102

Which element is a yellow, odorless solid at room temperature?

#103

Write the equation that represents beta-decay.

#104

Aspartic acid is a representative amino acid. Write the formula for aspartic acid.

CHEMISTRY QUICKIES

#105

What is the difference between a voltaic cell and an electrolytic cell?

#106

Give the name for the process of converting a combined metal ion in a mineral into a free metal? An example of a reaction of this process would be the following:
$Cu^{2+} + 2e^{-} \rightarrow Cu$

#107

What is the job of a limiting reactant in terms of products?

CHEMISTRY QUICKIES

#108

How many repeating units compose a polymer?

#109

Compound A has this structure, $R-\overset{O}{\overset{\|}{C}}-OH$.

Compound B has this structure, $R-\overset{O}{\overset{\|}{C}}-OR$. Which compound is an ester, and which is a carboxylic acid?

#110

What purpose does sodium chloride serve in making homemade ice cream?

CHEMISTRY QUICKIES

#110-S

SAFETY HINT:

Carefully read over every experiment before attempting to perform it.

CHEMISTRY QUICKIES

#111

How can corrosion of a metal object be prevented? Process?

#112

Name the acid found in milk?

#113

What gas does an athlete expel more rapidly than necessary when he hyperventilates? This upsets the carbonic acid equilibrium of the body.

CHEMISTRY QUICKIES

#114

If you were going to make nylon, what main chemical would you use?

#115

What is the fuel of choice for rockets, like the space shuttle?

#116

What does the second quantum number, ℓ, designate? How many values can ℓ have in a given energy level?

CHEMISTRY QUICKIES

#117

Why is cortisone, a compound synthesized by Percy Julian, so important?

#118

Why is it that if you touch a match to a pile of flour, it does not ignite, but flour floating in the air as dust increases the possibility of dust explosion?

#119

Which functional group characterizes an alcohol? (a) CH_3 (b) NH_3 (c) OH

CHEMISTRY QUICKIES

#120

Whose work led to atomic numbers being used for the Periodic Table instead of the use of atomic masses?

#121

The compound mannitol, $C_6H_8(OH)_6$, is used as a sweetener in some dietic foods. What is the mass of carbon in this compound?

#122

What is another name for the Group IIA elements?

CHEMISTRY QUICKIES

#123

What element is at the center of the chlorophyll molecule?

#124

In 1669 the systematic study of crystals began. Name the scientist who initiated this study?

#125

When dissolved gases in water reach a 124% saturation level, are humans threatened? These gases are primarily oxygen and nitrogen.

CHEMISTRY QUICKIES

#126

What is a buckyball?

#127

What percent of the air is nitrogen?

#128

What time of year does the ozone layer over the continent of Antartica thin to about half its normal thickness?

CHEMISTRY QUICKIES

#129

Name the most plentiful metal in the earth's crust.

#130

Osmosis has applications in the human body. When a red blood cell is surrounded by a solution of low concentration, what happens?

#131

In what kind of reaction do products react to form the original reactants?

CHEMISTRY QUICKIES

#132

How can increasing the concentration of a reactant increase the rate of a reaction?

#133

What suffix in the name of an enzyme would tell you this term is for an enzyme?

#134

What kind of substances are usually separated by gas chromatography?

CHEMISTRY QUICKIES

#135

Give the specific heat for $BaTiO_3$.

#136

What is the half-life of carbon-14?

#137

Which measurement contains three significant figures? (a) 0.004g (b) 0.040g (c) 0.0404g (d) 0.40g

CHEMISTRY QUICKIES

#138

When batteries in a car are recharged, some of the water molecules in the electrolyte are converted into gases at each electrode. Because the gases rise to the surface as bubbles, water has to be added to the batteries occasionally. One gas is at the cathode, and the other is at the anode. Name the gas at each electrode.

CHEMISTRY QUICKIES

#139

What is the meaning of molality?

#140

Rutherford and his colleagues aimed positively charged alpha particles at thin sheets of what element?

#141

If you have ever looked for an antacid in a pharmacy, you undoubtedly ran across Milk of Magnesia, which is a hydroxide compound. What element is a main component along with the hydroxide ion in Milk of Magnesia?

CHEMISTRY QUICKIES

#142

When energy is applied, many substances will break up or decompose. What type of reaction is this?

#143

At equal temperatures and equal pressures, equal volumes of gas contain how many molecules?

#144

When nearly perfect crystals are "doped" they turn into something quite useful. In to what are they turned?

CHEMISTRY QUICKIES

#145

State Raoult's Law.

#146

Because transition metals can be toxic to the body, physicians use polydentate ligands to solve this problem. How?

#147

How do you calculate the normality of a solution?

CHEMISTRY QUICKIES

#148

What is an ideal solution?

#149

How are acclelerators used to help understand nuclear structure?

#150

What property of neutrons and protons create a magnetic field?

CHEMISTRY QUICKIES

#151

If someone asked you to connect food and radiation sterilization, what would you say?

#152

What type of solution scatters light?

#153

During electrolysis, what happens to a moving ion when it reaches the electrode to which it is attracted?

CHEMISTRY QUICKIES

#154

Identify the element having this electronic configuration: $1S^22S^22P^63S^23P^64S^23D^8$

#155

List the elements that have characteristics of both metals and nonmetals.

#156

Identify the element having this configuration: $1S^22S^22P^6$

CHEMISTRY QUICKIES

#157

What is another name for resources that can be replenished through natural processes?

#158

If you do not eat enough, how will your body attain the energy it needs?

#159

What kind of bonding has an unusually strong attraction between a polar hydrogen atom and a strongly electron attracting atom like nitrogen or oxygen?

CHEMISTRY QUICKIES

#160

What kind of bonding involves sharing of electrons?

#161

What is the common name for sodium hydroxide?

#162

What is the formula for ozone?

CHEMISTRY QUICKIES

#163

Name three elements found in fertilizer.

#164

Name a pharmaceutical product used to neutralize excess stomach acid.

#165

When a tornado passes a region, is there a dramatic change in pressure?

CHEMISTRY QUICKIES

#166

MADAM C.J. WALKER

Madam C.J. Walker created a system of hair care products for Black women around the turn of the twentieth century and became America's first African American woman millionaire.

What element may have been the secret ingredient in her hair care products?

#167

Oxidation-reduction reactions involve either a loss or a gain of electrons. What is the oxidation number for an uncombined element in a chemical reaction?

#168

Which of the major air pollutants derives from lightning or bacterial action in soil?

#169

What color does $FeSO_4$ turn when KSCN is added to it?

CHEMISTRY QUICKIES

#170

What is the common name for Cu_2S (Not its representative chemical name)?

#171

What kind of gas or vapor is produced when a halide is oxidized?

#172

In the chemical industry, which substance is produced in largest quantity?

CHEMISTRY QUICKIES

#173

How close to a flame can you safely place hexane? What determines this distance?

#174

What is the characteristic color of iodine in the vapor form?

#175

Would you classify milk as largely a carbohydrate, fat, or protein type of food?

CHEMISTRY QUICKIES

#176

Narcotic analgesics have a medical value. What is that medical value?

#177

When hydrogen burns in oxygen a flame is produced. What is the characteristic color of this flame?

#178

Is the pH scale a measure of acidity or basicity?

CHEMISTRY QUICKIES

#179

Which has greater viscosity--honey or lemonade?

#180

What does compression of a gas do to temperature in a refrigerator?

#181

Which group of elements on the Periodic Table is known as "salt forming?"

CHEMISTRY QUICKIES

#182

What is the SI unit for mass?

#183

What is the symbol for the carboxylic acid group?

#184

What percent of your body's non-water mass consists of proteins?

#185

Why is Lise Meitner important to nuclear chemistry?

#186

How much energy does one gram of fat produce? Answer in terms of kilo Joules.

#187

What percent of the earth's total water is stored in the oceans?

CHEMISTRY QUICKIES

#188

How many liters of water containing liquids must a human take in daily to make up for bodily fluids lost daily?

#189

What is the nationality of renowned chemist, Percy Julian, who died in 1975? Julian headed Julian Laboratories, centering much of his work on steroids and hormones.

#190

When you add salt to ice, does it lower or raise the freezing point of ice below or above that of pure ice?

CHEMISTRY QUICKIES

#191

What is the purpose of a moderator in a nuclear reactor?

#192

Radium and hafnium are radioactive elements. Is the atomic number for radium higher than that of hafnium?

#193

What is the function of a salt bridge?

CHEMISTRY QUICKIES

#194

GARRETT MORGAN

During World War I, the U.S. Government converted the Morgan inhalator into the "gas mask." The undisputed value of the inhalator was demonstrated in July, 1916, through rescue operations of men trapped in a tunnel around Cleveland.

What year did Garrett Morgan die?

CHEMISTRY QUICKIES

#195

Will a metal conduct electricity if it is in the liquid state?

#196

Name the metal that is liquid at room temperature.

#197

What does it mean to say that an ion is reduced?

CHEMISTRY QUICKIES

#198

In 1980 the U.S. Department of Mint and Engraving replaced 97% of the copper in a penny with another metal. Which metal replaced copper in the penny?

#199

When sulfuric acid is added to water, the solution temperature rises quickly. To lower the risk of the solution boiling and spattering, you could immerse the mixing vessel in a material to control the solution temperature. Should the material for temperature control be (a) alcohol or (b) ice?

#200

Look at your Periodic Table. Carbon is atomic number 6. How many protons does carbon contain?

CHEMISTRY QUICKIES

#201

The atomic number of an element is 25. What is the element?

#202

Lloyd Hall held more than 25 patents for packaging methods and food processing, including recipes for curing salts and pickling solutions. What is the purpose of curing salts?

#203

At 25°C, the concentrations of hydrogen ions and hydroxide ions are equal in pure water, after the ions dissociate. What is that equal concentration?

CHEMISTRY QUICKIES

#204

Which of the following transition metals is used in the filaments of light bulbs: (a) gold (b) tungsten (c) cobalt?

#205

Which of the following metals is used in plumbing, electrical wiring, and coinage: (a) platinum (b) radium (c) xenon (d) copper?

#206

When tetrafluoroethane undergoes addition polymerization, it forms an important product that we find in some kitchen pots and other utensils. Name that product.

CHEMISTRY QUICKIES

#207

When vinyl chloride undergoes addition polymerization, it forms a product used to produce stereo records, drinking tumblers, telephones, combs, toys, and other common household plastics. Name that product.

#208

What color is copper sulfate solution?

#209

Two fine silk threads are cemented together by sericin, a gum that is produced by glands in the jaw of the silkworm. Sericin needs to be removed during the processing of silk by soaking cocoons in basins of a hot liquid. Is that liquid (a) milk (b) hydrochloric acid (c) water or (d) perfume?

CHEMISTRY QUICKIES

#210

Suppose you run an analysis on a solution to separate Ag ions from Ba ions. You add sodium chloride (NaCl). One of the ions forms a chloride and precipitates as a white solid, but the other remains in solution although it also forms a chloride. Which ion precipitates as a white solid?

#211

If H_2S (hydrogen sulfide) is added to an acidic solution containing only cadmium and nickel, which element would precipitate, since it forms one of the most insoluble sulfides in an acidic environment?

#212

Since silver chloride is insoluble in water, will it dissolve in aqueous ammonia?

CHEMISTRY QUICKIES

#213

Tooth enamel is composed of the mineral hydroxyapatite, $Ca_5(PO_4)_3OH$, which dissolves in acidic saliva. Affected teeth can be bathed in a fluoride solution, and the F^- replaces the OH^-. Do you think fluoride provides a weak base or a weak acid environment to bring about stronger teeth?

#214

A mixture of two concentrated acids is called aqua regia. Name the two acids used for this mixture.

#215

You are in the laboratory identifying elements by performing the Flame Test on salts. You observe a violet color in the flame, after testing a particular salt. Predict the element giving off this color.

#216

You are in the laboratory performing the Flame Test. For copper, what is the identifying color in the flame?

#217

While running an analysis for lead, you add potassium chromate to a solution of $Pb(NO_3)_2$, lead (II) nitrate. A precipitate forms. What color is this precipitate?

#218

When iron metal is added to a solution of Cu ions, the copper is reduced. What happens to the iron?

CHEMISTRY QUICKIES

#219

What does it mean to say that an ion is oxidized?

#220

Positive ions in solution are attracted to the cathode. Give another name for positive ions.

#221

Negative ions in solution are attracted to the anode. Give another name for negative ions.

CHEMISTRY QUICKIES

#222

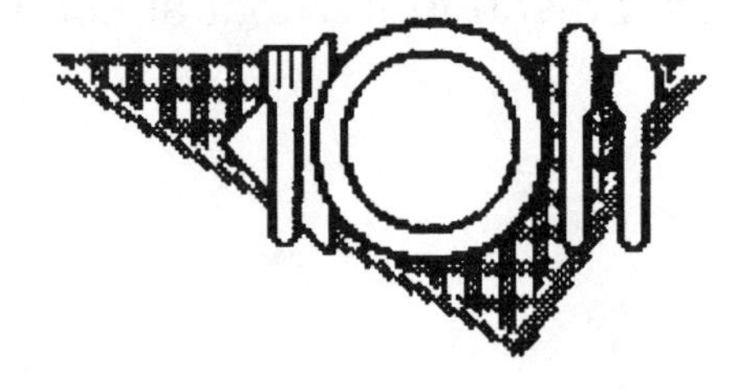

SAFETY HINT:

Eating and drinking in the laboratory are strictly forbidden.

#223

Pure carbon burning in oxygen releases heat at a rate of 94Kcal per mole of carbon. Is this an endothermic reaction or an exothermic reaction. Why?

#224

Cynthia performed an experiment to determine the heat of fusion of ice. She uses a 25 mL volume of hot water. Calculate the heat lost by the hot water. The temperature change for the hot water was 50.0°C. Answer in Joules.

#225

Cynthia wanted to determine the molar heat of fusion of ice. She recorded that 20.0 grams of ice melted; molar mass of ice equals 18 g/mol. Previous calculation found that the heat absorbed by the water was 5225 Joules. She used the equation:

Molar heat of fusion of ice =

(Heat lost by water ÷ Mass of melted ice) x Molar Mass of ice

CHEMISTRY QUICKIES

#226

Omar carried out a Boyle's Law experiment. Afterward, he plotted a graph of volume versus pressure of the confined gas. He arrived at a smooth curve through the plotted points. Name the shape of the curve he obtained and explain the indications of the curve.

#227

A Chemistry class produced orthorhombic sulfur by adding a small pea size quantity of sulfur to a tiny volume of vegetable oil in a 50-mL beaker and heating the mixture over a low flame. When they examined the resulting product under a microscope, did they find it to resemble (a) small needles or (b) blocky crystals?

#228

With isotopes of an element, what value changes--atomic number or atomic masses?

CHEMISTRY QUICKIES

#229

Several allotropes of sulfur can be produced in the laboratory. One allotrope, plastic sulfur, has no crystalline form when it is first produced. Give another name for "no crystalline form."

#230

The boiling point of sulfur is 444°C. What color is sulfur at its boiling point?

#231

Traci and Darci wanted to calculate the solubility product for lead chloride, $PbCl_2$. The equation for the reaction is the following:

$PbCl_2(s) \rightleftarrows Pb^{++}(aq) + 2Cl^{-}(aq)$

Write the solubility product expression for the above reaction.

CHEMISTRY QUICKIES

#232

One technique for separating substances in a mixture is chromatography. Although this method can be used for all kinds of mixtures, what particular physical property would allow easy detection of components in the mixture?

#233

Would you use a platform balance if great accuracy in measurement is required?

#234

Which of the following is not a physical property?

(a) Iron filings are insoluble in water.
(b) Sucrose is not attracted by a magnet.
(c) Mary measured a length of two feet for ribbon.
(d) Upon burning, magnesium metal leaves a white residue.

CHEMISTRY QUICKIES

#235

Identify the scientist who wrote the book, THE NATURE OF THE CHEMICAL BOND, was awarded the 1954 Nobel Prize for Chemistry, and received the 1962 Nobel Prize for Peace. This scientist also believed that vitamin C contains health benefits for cancer as well as the common cold.

#236

The following drawing shows a molecule containing one lone pair of electrons and three bonding pairs of electrons. Name this molecule:

$$\begin{array}{c} \cdot\cdot \\ H:\!N\!:H \\ \cdot\cdot \\ H \end{array}$$

#237

Name the simplest hydrocarbon compound.

CHEMISTRY QUICKIES

#238

Do water molecules attract each other because they possess (a) positive and positive ends (b) positive and negative ends or (c) negative and negative ends?

#239

Would you expect a polar sovent to dissolve a nonpolar substance?

#240

In the laboratory, you test a substance and find that it is a brittle solid with a high melting point. Would you classify this substance as covalent or ionic?

#241

If you could count the number of atoms in 64 grams of copper or the number of atoms in 52 grams of chromium, they would equal. What is the number of atoms in 1 mole of a sample of any element?

#242

Elizabeth Marie wanted to know the mass of water to be weighed to obtain 3.6 moles of H_2O. Solve for her.

#243

Jo Lingh set up a ring with double buret clamps and two burets. Using appropriate procedures, he prepared a standard acid solution and set out to standardize a solution of base. For this mixing of a measured volume of standard solution with an unknown solution, Jo Lingh engaged in what laboratory technique?

CHEMISTRY QUICKIES

#244

Endpoint in a titration results when you have added the necessary amount of base to react with the amount of acid given. What can you drop into your solution to signal when the end-point is reached?

#245

DNA and RNA differ in their structures. How?

#246

Bromothymol blue is an indicator that changes color from yellow to blue close to a pH of 7. If you were titrating sodium hydroxide and hydro-chloric acid, would bromothymol blue be an appropriate indicator? Why?

CHEMISTRY QUICKIES

#247

Which body process breaks down large colloidal particles before they can pass through the intestinal lining and enter the circulatory system?

#248

Excess salts, water, and waste products of the body are removed by the membranes of the kidneys through dialysis. The major waste product, urea, forms a substance in the bladder that is excreted. Name the substance formed.

#249

(a) $CH_3-\overset{\displaystyle CH_3}{\underset{\displaystyle CH_3}{\overset{|}{\underset{|}{C}}}}-CH_3$ and

(b) CH_3-CH_2-CH_2-CH_2-CH_3 are called structural isomers. Give the molecular formula for structures (a) and (b) shown above.

#250

SAFETY HINT:

Wear safety goggles at all times during a laboratory experiment.

CHEMISTRY QUICKIES

#251

Which of the following is not true for an Arrhenius acid?
(a) Forms hydrogen ions in aqueous solutions.
(b) Turns blue litmus red.
(c) Is a nonelectrolyte.
(d) Tastes sour.

#252

Which of the following is true for an Arrhenius base?
(a) Tastes sour.
(b) Forms hydrogen ions in aqueous solutions.
(c) Produces OH^- (hydroxide ions) in aqueous solutions.

#253

CH3-CH2-CH2-Br is named 1-bromopropane or propyl bromide. It belongs to a class of compounds called alkyl halides. Name the halogen atom attached to the alkyl group.

CHEMISTRY QUICKIES

#254

DDT is a chlorinated hydrocarbon and is nonbiodegradable. If absorbed by the human body, it is poisonous. Name the fly that DDT controled in earlier years before scientists became aware of its dangers.

#255

An enzyme in saliva breaks apart bonds in starch. Name that enzyme.

#256

Test to see whether bread, pasta, crackers, or other foods contain starch by dropping a little iodine onto a tiny amount of each sample. If starch is present, what color will the sample turn?

CHEMISTRY QUICKIES

#257

A high level of glucose in the urine is a symptom of which disease?

#258

"Black Gold" is another name for what substance?

#259

Soaps have a hydrocarbon end that attracts fats and grease. This is the nonpolar end. The polar end is water-loving and permits the soap to dissolve in water. If you described the polar end, would you label it (a) hydrophilic or (b) hydrophobic?

CHEMISTRY QUICKIES

#260

Triglycerides are present in our diets. What are the common names given to triglycerides?

#261

Skin care products often include alpha hydroxy acids, which are said to soften skin cells. Lactic acid, present in sour milk, is an alpha hydroxy acid. Some reports tell of Cleopatra bathing in sour milk to derive beautiful skin. Name two other alpha hydroxy acids.

#262

An acid found in the sting of a bee irritates the skin. Name that acid.

CHEMISTRY QUICKIES

#263

Give the names of the fruit acids (alpha hydroxy acids) found in the following fruits: grapes, apples, lemons, sugar cane.

#264

An acid is the oxidation product of the ethanol of wines and apple cider. Name that acid.

#265

Name the acid that gives the foul odor to rancid butter.

CHEMISTRY QUICKIES

#266

The addition of what molecule would convert aldehydes and ketones into alcohols?

#267

Methyl ethyl ketone (MEK) is the commercial name for butanone. How is MEK used?

#268

-273°C is the coldest temperature possible. Convert this into Kelvin.

#269

310K is normal body temperature. Convert this into Celsius temperature.

#270

Using an oxidation number charge chart, predict the formula for a compound formed from the union of these two ions: Potassium and oxygen.

#271

Using an oxidation number charge chart, predict the formula for a compound formed from the union of aluminum and sulfur.

CHEMISTRY QUICKIES

#272

When two pairs of electrons are shared, what type of bond is formed?

#273

When three pairs of electrons are shared, what type of bond is formed?

#274

Write the electron dot formula for calcium.

CHEMISTRY QUICKIES

#275

Will the reaction of a metal hydroxide with an acidic oxide yield H_2O as one of its products?

#276

In the oxidation-reduction reaction of potassium permanganate and hydrochloric acid, chlorine is one of the products. Chlorine is soluble in water. Would you collect the chlorine by water displacement or by air displacement?

#277

Perhaps the earliest concept of the atom as indivisible small particles came from a Greek philosopher of the era 460-362 B.C. Was that philosopher Democritus or Aristotle?

CHEMISTRY QUICKIES

#278

JOHN DALTON

John Dalton, after studying the experimental observations of others, proposed an atomic theory. The theory said that all matter is composed of atoms and that all atoms of the same element are identical. It also said that atoms of different elements are not alike.

What did Dalton's Atomic Theory say about the ratio of atoms as they form compounds?

CHEMISTRY QUICKIES

#279

If you mixed 0.50 grams of sulfur with 0.50 grams of iron filings, would you expect the sulfur or the iron filings to lose identity?

#280

Who is credited with being the first to prepare oxygen?

#281

Which gas can be prepared by reacting an active metal, such as zinc, with an acid, such as sulfuric? This gas has the lowest atomic mass and the simplest structure.

CHEMISTRY QUICKIES

#282

If a space is already occupied by a gas, will another gas be able to pass through that space simultaneously?

#283

You're performing a lab. You add concentrated sulfuric acid to sucrose. After the sugar breaks down, a porous, hard black substance remains. Name that substance.

#284

What is the hydroxide ion concentration, OH^-, in an aqueous solution at 25°C in which the hydronium ion concentration, H_3O, is 10^{-2}M?

CHEMISTRY QUICKIES

#285

What is the hydroxide ion concentration, $[OH^-]$, in an aqueous solution at 25°C in which the hydronium ion concentration, $[H_3O^+]$, is 10^{-5}M?

#286

Which of these is a saturated hydrocarbon?
(a) C_2H_2 (b) C_5H_{12}

#287

Benzene's molecular formula is C_6H_6. What is its empirical formula?

#288

How do you represent normality?

#289

Can the major food crops of the world utilize nitrogen in the air directly?

#290

How does the addition of a nonvolatile solute affect the freezing point of a solution?

CHEMISTRY QUICKIES

#291

True or false? The most active metals have the lowest electronegativities.

#292

In nuclear magnetic resonance, protons and neutrons produce motion which creates a magnetic field. What happens if 2 neutrons pair?

#293

Billie's chemistry class produced a cool bluish green light by oxidizing luminol. The same kind of reaction allows fireflies to flash light during a summer's evening as they release energy to attract mates. Emergency light sticks operate from a similar principle. What kind of reactions are involved in these displays of chemiluminescence.

CHEMISTRY QUICKIES

#294

Lanthanides, inner transition metals on the Periodic Table, are mined chiefly in California. Two lanthanides are useful in making lenses of welder's eye goggles and in the glass of television screens, because they tend to absorb intense light. Name these two lanthanides.

#295

Particularly during World War I a smokeless gunpowder was used because it nearly quadrupled the energy released from black gunpowder. Smokeless gunpowder has the chemical name of cellulose nitrate. What was the common name?

#296

What hardness of $CaCO_3$ needs to be present in hard-water to make it an economical solution to soften water?

#297

Perhaps you have made slime in a chemistry laboratory. Slime is a polymer gel that is fun to play with, moving it around with your stirring rod, as you watch it wiggle and flatten. What is the main constituent of slime?

#298

Calculate the formula mass for CsF.

#299

Who published the first useful Periodic Table?

CHEMISTRY QUICKIES

#300

During the Industrial Revolution, many city children with rickets were cured after being sent to the country. Why?

#301

What is Avogadro's number in numerical form?

#302

Which element is located in a group of metals but has properties of gases?

CHEMISTRY QUICKIES

#303

Who is Henry Mosley?

#304

Which Gas Law states that the volume of a dry gas is inversely proportional to its pressure, if the temperature is held constant?

#305

When the volume of a dry gas increases and the pressure is held constant, would you expect its temperature to decrease or increase?

CHEMISTRY QUICKIES

#306

DR. SAMUEL MASSIE

Dr. Samuel Massie received the 1994 Jack Flack Norris Award for Outstanding Achievement in the teaching of Chemistry from the Northeastern Section of the American Chemical Society. In 1986, he received the first Lifetime Achievement Award for Sustained Excellence in Science and Technology and Community Service. His work on drugs that fight against gonorrhea, malaria, and meningitis brought him a Citation from the Army. Having earned his first college degree at the age of 18, he later became the first African American professor hired by the Naval Academy.

In what year was Dr. Massie hired by the Naval Academy?

REFERENCE
Guide To Topic Items

- -
- -

The following guide will help you locate items of interest by special topics. Quickie numbers indicate location, rather than page number.

CHEMISTRY QUICKIES

MOLE

#16, #241, #242

MIXED CONCEPTS

#22, #25, #33, #62, #67, #75, #146, #157, #202, #235

NUCLEAR CHEMISTRY

#17, #30, #31, #60, #66, #70, #103, #136, #149, #150, #185, #191, #192

ORGANIC COMPOUNDS

#15, #26, #27, #28, #42, #44, #49, #53, #58, #59, #61, #63, #69, #87, #92, #93, #96, #97, #104, #108, #109, #114, #117, #119, #133, #163, #173, #175, #183, #189, #206, #207, #209, #237, #245, #249

OXIDATION-REDUCTION

#72, #106, #167, #197, #218, #219, #220, #221

PERIODIC TABLE & ELEMENTS

#12, #13, #19, #28, #37, #40, #45, #46, #47, #51, #57, #73, #78, #95, #93, #102, #115, #122, #123, #126, #127, #128, #129, #155, #174, #177, #181, #195, #196, #200, #204, #205

PH

#5, #6, #85, #178, #246

REACTION RATE & CHEMICAL EQUILIBRIUM

#131, #132, #203, #231

CHEMISTRY QUICKIES

SOLIDS & LIQUIDS
#124, #144, #227, #229

SOLUTIONS

#48, #54, #139, #147, #148, #152, #239

WATER

#23, #24, #33, #55, #187, #238

CHEMISTRY QUICKIES
ANSWERS

1. True.
2. True.
3. Aluminum.
4. Carbon dioxide.
5. The pH of blood needs to range from 7.35 to 7.45.
6. A system of buffers helps maintain blood pH. The blood keeps an equilibrium between its carbonic acid and bicarbonate ions which neutralizes any acid or base that adds to the blood. This allows the pH to stay at its normal level.
7. $CaCl_2$
8. Vinegar is a 5% aqueous solution of acetic acid, made by fermenting fruit or grain.
9. Lime.
#10. Expansion is the property of water that causes sidewalks to crack and frozen water pipes to break.

#11. The formula mass in grams for sodium hydroxide, NaOH, is 40.0 grams.
#12. The symbol for bromine is Br.
#13. The symbol for americium is Am.
#14. Control the height of the hot-air balloon over ground by varying the temperature of air inside the balloon.

CHEMISTRY QUICKIES
ANSWERS

#15. The formula for beeswax is $C_{15}H_{31}COOC_{30}H_{61}$.

#16. Molarity is the single term that expresses the quantity of moles per volume.

#17. Transmutation is the conversion of one element into another.

#18. Dry ice is a form of carbon dioxide.

#19. The Periodic Table contains 70% metals, more metals than nonmetals.

#20. There is no change in temperature as ice melts.

#21. The common name for calcium hydroxide is slaked lime.

#22. HBO is a chemistry term meaning Hyperbaric oxygen. Hyperbaric oxygen chambers are used to treat premature infants, heart attack patients, and sports injuries. Oxygen is dissolved in the blood stream in larger amounts through this process.

#23. The cations in hard water which are attracted to the resin are calcium and magnesium.

#24. "Frothy" water is a mixture of air and water.

#25. George Washington Carver was associated with Tuskegee Institute, Iowa State University, and Simpson College.

CHEMISTRY QUICKIES

ANSWERS

#26. Complete protein contains all 8 essential amino acids.

#27. Elastomers are substances that can be deformed by an outside force, but they can assume their original shape if the outside force is removed.

#28. No.

#29. The rubber ball you might play with in bouncing or throwing is an elastomer.

#30. New York and South Carolina.

#31. They absorb neutrons efficiently.

#32. NH_4Cl.

#33. You know Iron (III) Thiocyanate by its red color.

#34. Paper is made from tree pulp.

#35. Diatomic molecule.

#36. 2 different elements make up a binary compound.

#37. Chlorofluorocarbons.

#38. Retinal takes the all-trans form.

#39. Yeast cells contain enzymes.

#40. The nervous system.

#41. Salt.

#42. Styrene butadiene rubber, sometimes called SBR, causes the large bubbles because it gives a stretchiness to the substance.

CHEMISTRY QUICKIES
ANSWERS

#43. Acetic acid and lactic acid. These acids are excreted when a group of bacteria metabolizes maltose.

#44. Oxygen can attack the double bonds in unsaturated fatty acids, because these bonds are highly reactive. Flavors and odors develop which are unpleasant. When unsaturated fats are left alone, food may spoil.

#45. Amethyst.

#46. China clay or kaolin has the chemical name aluminum silicate hydroxide.

#47. Clay is composed of silicon, oxygen, and aluminum, along with magnesium, sodium, and potassium ions.

#48. Alkaline solution.

#49. Ascorbic acid.

#50. 132.9

#51. Period. Group.

#52. Noble gases.

#53. Enzymes.

#54. Supersaturated.

#55. When there are no substances present that will form precipitates with soap, water is "softened."

#56. In 1911.

#57. Nitrogen.

CHEMISTRY QUICKIES
ANSWERS

#58. Albumin.
#59. Pepsin is an enzyme.
#60. The Savannah River Plant in South Carolina is the world's largest vitrification plant.

#61. Hexane.
#62. Yes.
#63. Proteins.
#64. Meats, salt-processed foods, table salt; sodium is the mineral.
#65. Carbon dioxide.
#66. Po-218
#67. 0.06%
#68. Ammonia.
#69. Margarine and shortening.
#70. Derby.

#71. 100°C
#72. Eventually it coats both electrodes and reduces their ability to produce current.
#73. Chromium impurities in the crystalline structure of emeralds and rubies are responsible for their colors.
#74. 540 calories.
#75. Jams are made from the entire fruit and jellies are made from the fruit juice.

CHEMISTRY QUICKIES
ANSWERS

#76. Because of hydrogen bonding, water molecules in ice are farther apart than in liquid water.
#77. Formic acid, $HCHO_2$.
#78. Corn.
#79. 8.26×10^{-10}
#80. Lactic acid.

#81. $Pb(CH_3COO)_2(aq) + K_2CrO_4(aq) \rightarrow PbCrO_4(cr) + 2KCH_3COO\ (aq)$
#82. $C \equiv C$
#83. Hydrochloric acid is found in gastric juices.
#84. Ammonium cyanate.
#85. A buffer contains an acid that can react with added base and it also contains a base that can react with any added acid.
#86. Charles Martin Hall.
#87. The body can not store any proteins.
#88. Never!
#89. The hottest tea floats at the top.
#90. No. It cools faster in the freezer.

#91. Pigs have no sweat glands, and this condition prevents them from cooling by evaporation of perspiration.
#92. An ester forms when an alcohol reacts with an acid.
#93. Octyl acetate.

CHEMISTRY QUICKIES

ANSWERS

#94. In the peroxide structure, oxygen has an oxidation number of 1^-.

#95. White phosphorous forms a vapor readily at room temperature and catches fire easily. It is volatile.

#96. Pentyl acetate (Pentyl ethanoate).

#97. Amyl butyrate.

#98. The number of neutrons.

#99. Standard temperature and pressure.

#100. Benzene is changed into catechol.

#101. Ideal gases.

#102. Sulfur.

#103. 1N → 1P + 0e + antineutrino

#104. $- OOC\text{-}CH\text{-}CH_2\text{-}COOH$ (with NH_3^+ attached to CH)

#105. In electrolytic cells, electric energy is used to cause chemical changes. In voltaic cells, chemical energy converts into electrical energy.

#106. Reduction.

#107. Limiting reactants limit the amount of products that can be formed.

#108. Polymers consist of 500 to 20,000 or more units.

CHEMISTRY QUICKIES

ANSWERS

#109. $R\text{-}\overset{O}{\overset{\|}{C}}\text{-}OH$ is an acid.

$R\text{-}\overset{O}{\overset{\|}{C}}\text{-}OR$ is an ester.

#110. Sodium chloride lowers the freezing point of water and maintains a lower temperature while freezing.

#111. Corrosion of a metal object can be prevented through electrolysis.

#112. Lactic acid.

#113. CO_2

#114. Hexanedioic acid.

#115. Hydrogen.

#116. Quantum number, ℓ, designates sublevels within an energy level. There can be n values of ℓ.

#117. Cortisone is used in the treatment of arthritis.

#118. Flour floating in the air spreads across a greater surface area. When it combines with oxygen, the chance for a reaction increases. Grain elevator explosions, that you may have read about, come from flour dust reacting with oxygen of the air.

#119. The -OH- group characterizes an alcohol.

#120. Henry Mosley.

CHEMISTRY QUICKIES

ANSWERS

#121. 72 grams
#122. Alkaline earth metals.
#123. Magnesium.
#124. Nicholas Steno.
#125. No. Fish get bubble disease but humans appear not to be directly affected.
#126. "Buckyball" is Budkminsterfullerene. It is an allotrope of carbon.
#127. 78% of the air is nitrogen.
#128. During the southern hemisphere winter.
#129. Aluminum is the most plentiful metal in the earth's crust.
#130. Fluid will flow into the cell and cause it to burst.

#131. Reversible reactions.
#132. Collisions are increased and this increases reaction rate.
#133. The suffix - "ase."
#134. Volatile liquids and mixtures are usually separated by gas chromatography.
#135. According to reference manuals, the specific heat of $BaTiO_3$ is 0.79418 J/g.°C.
#136. The half-life of carbon-14 is 5730 years.
#137. 0.0404g contains three significant figures.

CHEMISTRY QUICKIES
ANSWERS

#138. Hydrogen at the cathode. Oxygen at the anode.
#139. Moles per kilogram of solvent.
#140. Gold.

#141. Magnesium.
#142. Decomposition.
#143. 6.02 X 10^{23} molecules.
#144. Semiconductors.
#145. The vapor pressure of a solution varies directly as the mole fraction of solvent.
#146. Polydentate ligands form complex compounds with the metals.
#147. Multiply the molarity of a solution by the total charge of the positive ions in the compound.
#148. An ideal solution is one in which all intermolecular attractions are the same.
#149. Accelerators are linear or circular devices used to increase the velocity of charged particles.
#150. The property of spin creates a magnetic field.

#151. It is a method of food preservation that prevents food spoilage.
#152. Colloid.
#153. The moving ion gains electrons or loses electrons.

CHEMISTRY QUICKIES

ANSWERS

#154. Nickel is atomic number 28.
#155. Metalloids.
#156. Neon.
#157. Natural resources.
#158. The body will use up stores of fat.
#159. hydrogen bonding.
#160. Covalent bonding.

#161. Lye
#162. O_3
#163. Nitrogen, phosporus, and potassium.
#164. Use antacid tablets.
#165. Pressure may drop from 760mm to 560mm when a tornado passes over a region.
#166. Sulfur.
#167. Zero.
#168. Nitrogen oxides.
#169. Red.
#170. Chalcocite.

#171. Halogen gas.
#172. Sulfuric acid.
#173. Hexane is flammable and should not be placed close to a flame.
#174. Purple or violet.
#175. Milk is a protein.
#176. Narcotic analgesics relieve pain.
#177. Blue.

CHEMISTRY QUICKIES
ANSWERS

#178. pH scale is a measure of acidity.
#179. Honey has a greater viscosity than lemonade.
#180. Raises it.

#181. Halogens.
#182. Gram.
#183. -COOH-
#184. 50%.
#185. Lise Meitner explained how nuclear fission occurs.
#186. 38 KJ
#187. 97%.
#188. 2.
#189. Percy Julian was an American.
#190. Adding salt to ice lowers the temperature below that of pure water.

#191. It slows down neutrons through collisions; and this does not absorb them.
#192. Yes. 88>72
#193. The function of a salt bridge is to complete the circuit between two separate solutions without mixing them.
#194. Garrett Morgan died in 1963.
#195. Yes. A metal will conduct electricity if it is in the liquid state.
#196. Mercury is the only common metal that is liquid at room temperature.

CHEMISTRY QUICKIES
ANSWERS

#197. To say that an ion is reduced means that a substance gains electrons.
#198. Zinc.
#199. Ice surrounding the mixing vessel would help control the solution temperature.
#200. Carbon contains 6 protons.

#201. Manganese has atomic number 25.
#202. Curing salts prevent bacterial growth in meat.
#203. 1.0 X 10^{-7} mol/L.
#204. Tungsten is used in the filaments of light bulbs.
#205. Copper is a transition metal used in plumbing, electrical wiring, and coinage.
#206. Teflon is the product formed when tetrafluoroethane undergoes addition polymerization.
#207. Polyvinyl chloride forms when vinyl chloride undergoes addition polymerization.
#208. Copper sulfate solution is blue.
#209. The liquid is water.
#210. The silver ion (Ag^+) precipitates as a white solid (silver chloride).

CHEMISTRY QUICKIES
ANSWERS

#211. Cadmium forms an orange sulfide that is insoluble.
#212. Yes; silver chloride dissolves in aqueous ammonia to form $Ag(NH_3)_2^+(aq)$ and $Cl^-(aq)$.
#213. Weak base.
#214. Aqua regia is made up of concentrated HCl (hydrochloric acid) and concentrated HNO_3 (Nitric acid).
#215. Potassium gives a violet color in the Flame Test.
#216. Bluish green is the identifying color for copper in the Flame Test.
#217. The precipitate is yellow $PbCrO_4$.
#218. The iron is oxidized to Fe^{2+}.
#219. An ion is oxidized if it loses electrons.
#220. Another name for positive ions is cations.

#221. Negative ions are also called anions.
#222. Safety Hint requiring no answer.
#223. This is an exothermic reaction, because heat is released during an exothermic reaction.

CHEMISTRY QUICKIES
ANSWERS

#224. 25 ML H_2O = 25g because density of H_2O = 1g/mL
Heat Lost = mass of H_2O X specific heat X temperature change = (25.0g) (4.18 J/g°C) (50.0°C) = 5225 Joules
**Significant digits have not been used.

#225. Molar heat of fusion =
5225 J/20.00g X 18.00g/mol =
4703 J/mol

#226. The shape of the curve obtained by Omar in the Boyle's Law experiment is hyperbola. An inverse relationship between volume and pressure is indicated.

#227. Blocky crystals.

#228. Atomic masses change with isotopes of an element, since the number of neutrons vary with each isotope.

#229. Amorphous.

#230. Dark red.

#231. Ksp ($PbCl_2$) = [Pb^{2+}] [Cl^-]^2

#232. Color. Chromos means "color."

#233. No. Only use platform balances when approximate mass measurements are needed.

#234. d

#235. Linus Pauling.

#236. Ammonia.

#237. Methane.

CHEMISTRY QUICKIES
ANSWERS

#238. b.

#239. Generally, polar solvents do not dissolve nonpolar substances.

#240. Ionic

#241. 6.02×10^{23} atoms

#242. Molar mass of H_2O = 18.0grams

$$\frac{3.6 \text{ mol } H_2O}{} \left| \frac{18.0 \text{ grams}}{1 \text{ mol } H_2O} \right.$$

$= 3.6 \times 18$

$= 64.8 \text{ g } H_2O$

#243. Titration.

#244. An acid-base indicator.

#245. DNA, a double helix, contains deoxyribose and thymine, cytosine, guanine, and adenine. RNA only has a single chain of nucleotides. Instead of deoxyribose, RNA contains ribose. While it also contains cytosine, guanine, and adenine, it does not contain thymine. Also, RNA contains uracil.

#246. Yes. NaOH and HCl would reach and endpoint at pH=7, since pH=7 is neutral. Acids and bases neutralize each other.

#247. Digestion.

#248. Urine.

#249. C_5H_{12}

#250. Safety Hint requiring no answer.

CHEMISTRY QUICKIES
ANSWERS

#251. C.
#252. C.
#253. Bromine.
#254. It controlled the tsetse fly that spread malaria, typhus, and sleeping sickness.
#255. Amylase.
#256. Blue-black.
#257. Diabetes mellitus.
#258. Petroleum.
#259. (a) Hydrophilic.
#260. Fats and oils.

#261. Tartaric acid, malic acid, citric acid, or glycolic.
#262. Formic acid.
#263. Grapes contain tartaric acid. Apples and grapes contain malic acid. Lemons contain citric acid. Sugar cane contains glycolic acid.
#264. Acetic acid.
#265. Butyric acid.
#266. Hydrogen, H_2
#267. MEK is used to remove paint.
#268. 0 Kelvin or 0 K. Absolute zero.
#269. 37°C. °C=K-273
=310K-273
=37°C
#270. K_2O

CHEMISTRY QUICKIES

ANSWERS

#271. Al_2S_3
#272. Double bond.
#273. Triple bond.
#274. Ca:
#275. Yes. H_2O and salt.
#276. Air displacement. Chlorine's solubility in water makes it difficult to collect by water displacement.
#277. Democritus.
#278. Atoms unite in simple ratios to form compounds.
#279. Neither. In a mixture, substances do not lose their identities.
#280. Joseph Priestly first prepared oxygen on August 1, 1774.

#281. Hydrogen.
#282. Yes. Gases are able to diffuse. Graham's Law of Diffusion.
#283. Carbon.
#284. $OH^- = 1.0 \times 10^{-12}$
$K_w = [H_3O^+][OH^-] = 1.0 \times 10^{-14}$
#285. $OH^- = 1.0 \times 10^{-9}$
#286. (b) C_5H_{12}, because the structural formula shows single bonds. Each carbon-carbon bond in a saturated hydrocarbon is a single bond.
#287. CH.

CHEMISTRY QUICKIES
ANSWERS

#288. Represent normality with a N.
#289. No.
#290. It lowers the vapor pressure of solvent and causes it to freeze at a lower temperature.

#291. True.
#292. Their spins will be opposite and their magnetic fields cancel each other.
#293. Oxidation-reduction reactions.
#294. Neodymium and praeseodymium.
#295. Guncotton.
#296. 200 ppm $CaCO_3$ 200 parts per million
#297. Polyvinyl alcohol.
#298. Formula mass for CsF is 152 grams.
#299. Demetri Mendeleyev.
#300. In the country children enjoyed more sunlight. Sunlight prevents rickets, a disease that causes bones to become soft and easily deformed.

#301. 6.02×10^{23}
#302. Hydrogen.
#303. Henry Mosley was an English physicist who was responsible for the Modern Periodic Table.
#304. Boyle's Law.
#305. Temperature would increase.
#306. Dr. Samuel Massie was hired by the Naval Academy in 1966.

Glossary of Key Terms

A

acid: A substance which produces hydrogen ions in an aqueous solution.
alpha emission: When alpha particles cause the parent nuclide to die.
alpha particles: A positively charged particle identical to the helium nucleus and emitted from radioactive nuclei.
anions: Negative ions. Particles in a negatively charged state.

C

catalyst: A substance that speeds up the rate of a chemical reaction, but it does not undergo a permanent change.
cations: Positive ions. Elements in a positively charged state.
chemical equilibrium: A state in which the forward and reverse reactions take place at the same rate, and at this time the concentrations of the reactants and products remain constant.
chemical reactions: The change of one or more substances into new substances, during which bonds are broken and new ones formed.
concentration: Amount of solute per given quantity of solvent or solution.
control rods: Rods in a nuclear reactor that absorb neutrons and thus control the rate of reaction.

Glossary of Key Terms

D

density: Mass per unit volume.

E

elastomer: A material that returns to its original shape after a force stretches it, bends it, compresses it, or otherwise distorts it.
electrolysis: A process in which current is forced through a cell producing a chemical reaction.
electrolytic cell: A device or chamber in which a chemical change takes place through application of electrical energy.
endothermic reaction: A chemical reaction in which energy is absorbed.
exothermic: A chemical reaction in which energy is given off.
enzyme: A specialized protein of a biochemical nature that catalyzes rates of reactions.

F

formula mass: The sum of the atomic masses of atoms in a compound, as represented by the formula.

Glossary Key of Terms

G

gas chromatography: An analytical technique for determining information on vapors using an inert carrier gas which distributes the vapor through a packed column.

H

half-time: The time it takes an element to decrease to half its original concentration. A radioisotope sample decays to half its size.
halide: A negatively charged ion of the halogen atom.
hard water: Water containing high concentrations of calcium and magnesium ions.
heat of fusion: The amount of heat needed to melt 1 gram of solid into a liquid.
hydrogenated oils: Vegetable oils or margarine. Hydrogen is added to double-double bonds in a controlled manner.

L

limiting reactant: The reactant that is used up first during a chemical reaction, thereby determining how much product is formed.

M

molecular formula: A formula showing the actual number of atoms of an element present in a molecule of a compound.

Glossary of Key Terms

M

moderator: In a nuclear reactor, substances which slow down neutrons.
molar heat of fusion: The heat needed for one mole of solid to liquefy.

N

normality: A measurement of concentration of a substance denoting its number of equivalents in one liter of solution.

O

osmosis: The movement of a solvent through a semipermeable membrane, from an area of high solvent concentration toward an area of lower solvent concentration.
oxidation number: The charge on an atom if the electrons in the bonds of a compound were assigned according to a set of established rules.

P

Periodic Table: The arrangement of elements into rows and columns, according to increasing atomic number; those with similar properties appear in a vertical column.
pico: A measurement of 1.0×10^{-9} units.

Glossary of Key Terms

P

pH: The negative log of the hydrogen ion (or hydronium ion) concentration.

Polymer: A large molecule formed from repeating units of monomers (small molecules). The monomers are covalently bonded to form the polymer.

polymerization: The process of joining many monomers (small molecules) together to form a very large molecule called a polymer.

protein: A biological polymer formed from the condensation reactions between amino acids, recognizable by the amide groups (NH_2) linking amino acids.

Q

quantum number: A number that describes an electron in an orbital.

R

reactants: Substances present at the start of a reaction, before a chemical change takes place.

resin: Nonvolatile organic semisolids or solids originating from plant discharges or from polymerization of simple molecules.

Glossary of Key Terms

S

salt bridge: Usually a u-tube, in a laboratory set-up containing a concentrated electrolyte solution, that connects half-cells in a voltaic cell. Ions pass from one half-cell to the other but are prevented from completely mixing with each other.
saturated: Hydrocarbons having single carbon-carbon bonds.
significant digits: Those digits in a measurement that are certain, plus one last digit which is presumed to be certain.
specific heat: The amount of heat needed to raise the temperature of one gram of a substance by one degree celsius.
soft water: Water that contains less than 120 ppm of $CaCO_3$.

T

transition metals: Elements in the central groups of the Periodic Table, having their highest energy electrons in a d sublevel.
transmutation: By the emission of radiation, the atom of one element is changed into the atom of another element.
triglycerides: Fats. Long chain fatty acids replace hydroxyl groups on a glycogen molecule.

Glossary of Key Terms

U

unsaturated: Hydrocarbons having one or more double or triple bonds.

V

viscosity: Resistance of a liquid to flow.
vitrification: A process for converting radioactive wastes into a solid. Glass is usually the solid.
voltaic cell: An electrochemical cell in which the chemical energy is changed into electrical energy.

CHEMISTRY QUICKIES

INDEX

A

B

INDEX

C

D

INDEX

E

F

G

INDEX

H

I

J

K

INDEX

L

M

N

INDEX

INDEX

S

T

INDEX

U

V

W

A SELECTED BIBLIOGRAPHY

The references listed below may serve as sources of more detailed information.

A Project Of The American Chemical Society. CHEMCOM. 2nd ed. Dubuqu, IA: Kendall/ Hunt Publishing Company, 1993.

Brady, James E. and John R. Holum. CHEMISTRY: THE STUDY OF MATTER AND ITS CHANGES. New York: John Wiley & Sons, Inc., 1993.

Kroschiwitz, Jacqueline I., Melvin Winokur, and A. Bryan Lees. CHEMISTRY: A FIRST COURSE. 3rd ed. Dubuque, IA: Wm. C. Brown, 1995.

Lemay, H. Eugene. CHEMISTRY: CONNECTIONS TO OUR CHANGING WORLD. Needham, Massachusetts: Prentice Hall, 1996.

Masterton, William L., and Cecile N. Hurley. CHEMISTRY: PRINCIPLES & REACTIONS. 2nd ed. Ft.Worth: Saunders College Publishing, 1993.

A SELECTED BIBLIOGRAPHY

Smoot, Robert C., Richard G. Smith, and Jack Price. CHEMISTRY: A MODERN COURSE. Columbus, Ohio: Merrill Publishing Company, 1995.

Timberlake, Karen C. CHEMISTRY. 6th ed. New York, NY: Harper Collins, 1996.

Tocci, Salvatore, and Claudia Viehland. CHEMISTRY: VISUALIZING MATTER. Austin: Holt Rinehart Winston, 1996.

Wilbraham, Antony C., et al. CHEMISTRY. Menlo Park, CA: Addison-Wesley, 1987.

Wistrom, Cheryl, John Phillips, and Victor Strozak. CHEMISTRY: CONCEPTS and APPLICATIONS. New York, NY: Glencoe, 1997.

Zumdahl, Steven S. CHEMISTRY. 3rd ed. Lexington, MA: D.C. Heath, 1993.

About the Author

Vivian W. Owens is the author of parent-helper books, PARENTING FOR EDUCATION (**PFE**) and CREATE A MATH ENVIRONMENT. Her first juvenile novel, NADANDA THE WORDMAKER, received a Writer's Digest award. Other books include juvenile novels, THE ROSEBUSH WITCH, I MET A GREAT LADY, and I MET A GREAT MAN. In addition to writing books, she teaches high school chemistry, writes a PFE newspaper column, and conducts PFE Workshops. A graduate of Tuskegee University (B.S.) and James Madison University (M.S.), she lives with her engineer husband, John, in Virginia. She is the mother of John, Shea, and April.

BOOKS by Vivian Owens

Mail Order
from
ESCHAR PUBLICATIONS

Make a collection of books
by Vivian Owens.
Order today and books will arrive
at your door within two to four weeks.

ESCHAR PUBLICATIONS
P.O. Box 1196
Waynesboro, VA 22980
Fax # (540) 942-3650

Please send me a free ESCHAR PUBLICATIONS brochure for home delivery.

Name______________________________

Address____________________________

City_______________State/Zip_________

Please send me a free ESCHAR PUBLICATIONS brochure for home delivery.

Name______________________________

Address____________________________

City_______________State/Zip_________